All About Antarctic Krill

Meg Greve

Creating Young Nonfiction Readers

EZ Readers lets children delve into nonfiction at beginning reading levels. Young readers are introduced to new concepts, facts, ideas, and vocabulary.

Tips for Reading Nonfiction with Beginning Readers

Talk about Nonfiction

Begin by explaining that nonfiction books give us information that is true. The book will be organized around a specific topic or idea, and we may learn new facts through reading.

Look at the Parts

Most nonfiction books have helpful features. Our *EZ Readers* include color photographs and graphic aids, a table of contents, a glossary, and an index. Share the purpose of these features with your reader.

Color Photos and Graphic Aids

A lot of information can be found by "reading" photos, charts, maps, and other graphic aids found within nonfiction texts. Help your reader learn more about the different ways information can be displayed.

Table of Contents

Located at the front of the book, this list shows the big ideas within the text and the page numbers where they can be found.

Glossary

Located at the back of the book, the glossary defines key words and phrases that are related to the topic. These words and phrases can be found in the text in colored type.

Index

Located at the back of the book, an index is an alphabetical list of topics and the page numbers where they can be found.

With a little help and guidance about reading nonfiction, you can feel good about introducing a young reader to the world of *EZ Readers* nonfiction books.

Mitchell Lane
PUBLISHERS

2001 SW 31st Avenue
Hallandale, FL 33009
mitchelllanepub.com

First Edition, 2026.

Author: Meg Greve
Designer: Rhea Magaro
Editor: Kim Thompson

Names/credits:
Title: All about Antarctic Krill / by Meg Greve
Description: Hallandale, FL :
Mitchell Lane Publishers, [2026]

Series: Animals Around the World
Library bound ISBN: 979-8-89260-517-5
eBook ISBN: 979-8-89260-524-3

Library of Congress Control Number: 2025933773

EZ Readers is an imprint of
Mitchell Lane Publishers

PHOTO CREDITS
Alamy: Auscape International Pty Ltd, cover, 1, 5; robertharding, 12; SeaTops, 14; Nature Picture Library, 19; imageBROKER.com GmbH & Co. KG, 21; Dreamstime: Izanbar, 10; Shutterstock: Tarpan, 8, 16; Mozgova, 16; Mario_Hoppmann, 16; Tory Kallman, 16; Blake Karas, 16; deer boy, 22.

Contents

Whales need them. Penguins need them. You could say the whole world needs them! They are tiny animals called krill. Trillions live in the Southern Ocean around the icy continent of Antarctica.

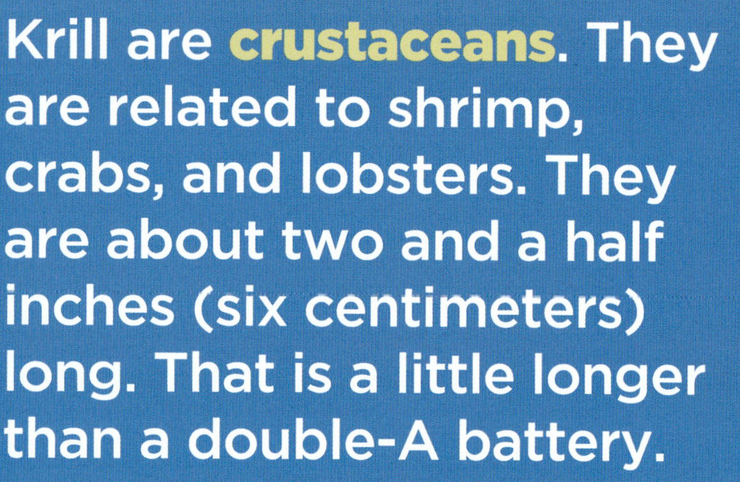

Krill are **crustaceans**. They are related to shrimp, crabs, and lobsters. They are about two and a half inches (six centimeters) long. That is a little longer than a double-A battery.

Krill have hard outer shells called **exoskeletons**. To swim, a krill moves its tail and 10 legs. Up to 16 more legs are used to filter water and trap food. Krill eat tiny plants and **algae** found under floating ice.

10

Krill have light red bodies that are **transparent**. Krill are also **bioluminescent**. That means they glow!

Millions of krill swim together in huge swarms. Their swarms are so big that they can be seen from outer space.

During the day, krill stay deep in the water. At night, they come up to eat. As they move around, krill provide food for animals in all parts of the ocean.

Food Chain

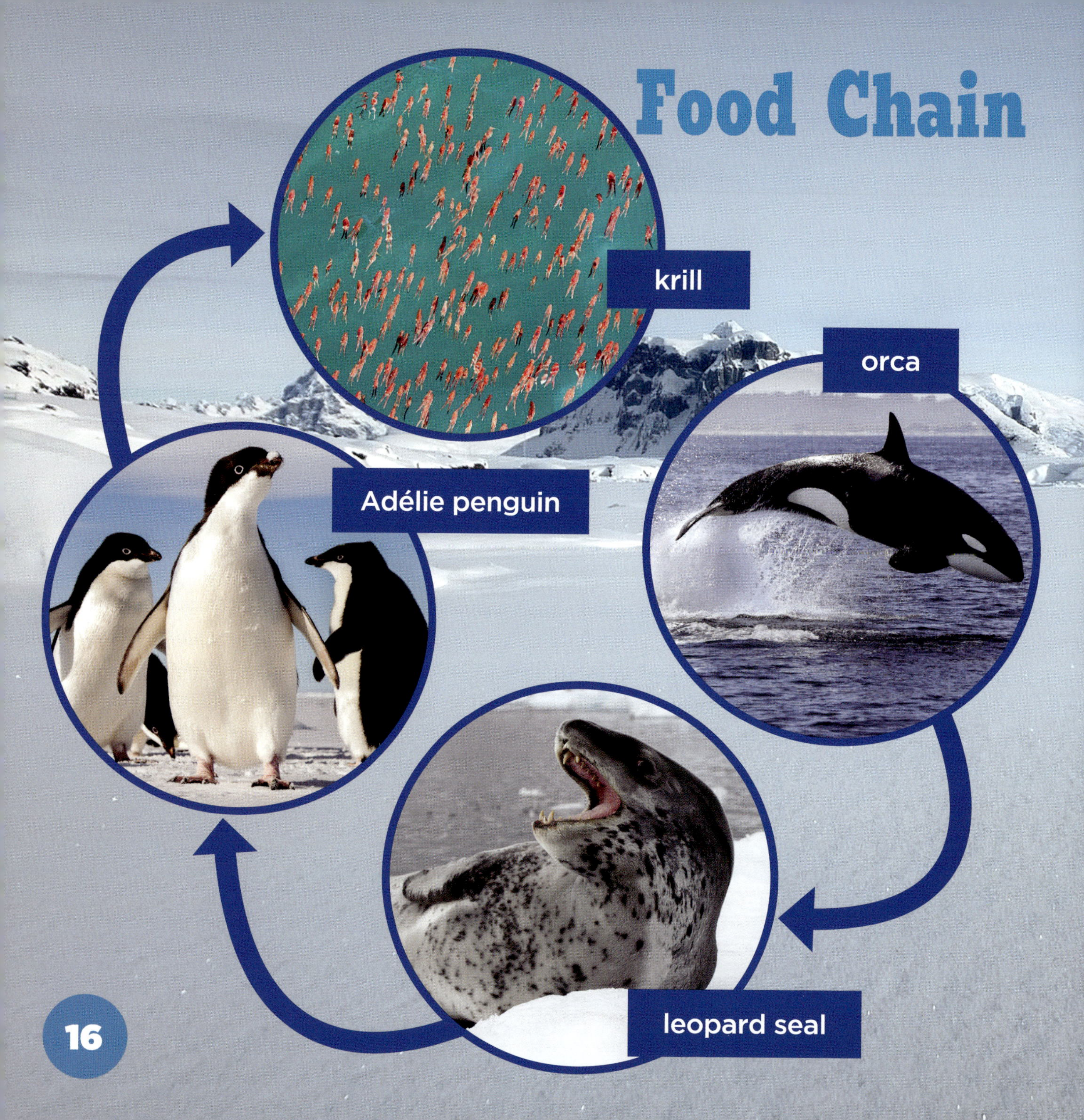

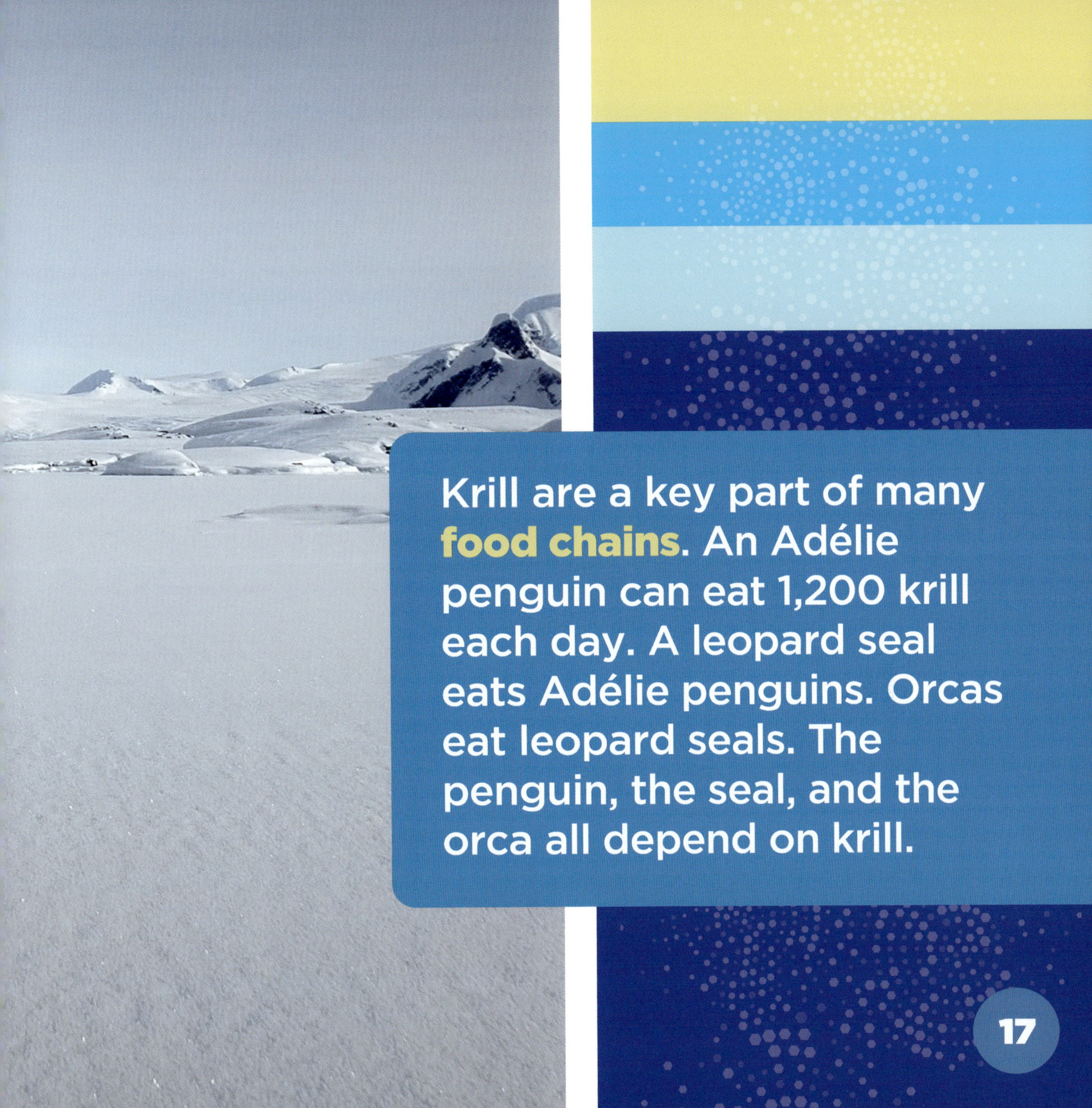

Krill are a key part of many **food chains**. An Adélie penguin can eat 1,200 krill each day. A leopard seal eats Adélie penguins. Orcas eat leopard seals. The penguin, the seal, and the orca all depend on krill.

Squid, fish, and whales also eat krill. Blue whales are the largest animals on Earth. They mostly eat krill. They can eat more than three million krill in one day.

The health of many animals depends on krill. These tiny creatures are important to Antarctica. They are important to oceans and living things around the world.

Where Do Krill Live?

Krill live in oceans around the world. Many swim in the Southern Ocean around Antarctica. They like to stay close to the ice. They eat the algae that grows under it.

Interesting Facts

- When krill do not have food to eat, they can shrink their bodies.
- The area of ocean filled by Antarctic krill is as large as the continent of Africa.
- All the krill in the world weigh about as much as all the people in the world!
- The name *krill* comes from Norway. It means "whale food."

Parts of a Krill

antennae
A krill has two antennae on its head.

exoskeleton
Krill shed these hard outer shells as they grow and throughout their lives.

eyes
The older a krill is, the larger its eyes are.

gills
Like fish, krill have gills that let them breathe underwater.

legs
Krill have six to eight pairs of legs for feeding and five pairs for swimming.

light organs
A krill has body parts under its belly that light up. The light may help krill communicate with each other. It may help them blend in with the environment.

stomach
A krill's stomach turns bright green after it eats ocean plants.

Glossary

algae (AL-jee)
Small plants without roots or stems that grow in water

bioluminescent (bye-oh-loo-muh-NE-suhnt)
Able to produce light as a result of chemicals in the body

crustaceans (kruh-STAY-shuhnz)
Sea animals that have hard outer skeletons, such as crabs, lobsters, and shrimp

exoskeletons (ek-soh-SKE-luh-tuhnz)
Hard outer shells that support some animals that do not have bones

food chains (food chaynz)
Diagrams that show the order in which living things eat other living things in an environment

transparent (trans-PAIR-uhnt)
Clear; see-through

Index

Further Reading

Lilley, Matt. *Good Eating: The Short Life of Krill.* Tilbury House Publishers, 2022.

Markovics, Joyce. *Shrimp and Krill (Lights On! Animals That Glow).* Cherry Lake Publishing, 2022.

On the Internet

Mongabay Kids: Thrill to the Krill
kids.mongabay.com/thrill-to-the-krill
Read about krill and watch a video of krill swimming.

World Wildlife Fund: Top 10 Facts about Antarctic Krill
www.wwf.org.uk/learn/fascinating-facts/antarctic-krill
Learn fascinating facts about these small sea creatures.